NUMERICAL METHODS PRACTICAL

NUMERICAL METHODS PRACTICAL

[For Honours and postgraduate students of all Indian universities]

Shyamapada Dutta
Lecturer in Mathematics
Bankura Sammilani College, Bankura

Harisankar Mallick
Lecturer in Mathematics
Ramakrishna Mission Vidyapith, Purulia

New Central Book Agency (P) Ltd
LONDON
HYDERABAD ERNAKULAM BHUBANESWAR
NEW DELHI KOLKATA PUNE GUWAHATI

NCBA

REGD OFFICE
8/1 Chintamoni Das Lane, Kolkata 700 009, India
email: ncbapvtltd@eth.net

OVERSEAS
NCBA (UK) Ltd, 149 Park Avenue North, Northampton, NN3 2HY, UK
email: ncbauk@yahoo.com

EXPORT
NCBA Exports Pvt. Ltd, 212 Shahpur Jat, New Delhi 110 049
email: ncbaexports@ncbapvtltd.com

BRANCHES

212 Shahpur Jat
New Delhi 110 049
email: ncbadel@ncbapvtltd.com

Shop Nos. 3 & 4, Vinayak Towers
681/B Budhwar Peth, Appa Balwant Chowk
Pune 411 002
email: ncbapune@ncbapvtltd.com

Shop No. 15, Bharati Towers, Ground Floor
Forest Park, **Bhubaneswar** 751 001
email: ncbabub@ncbapvtltd.com

House No. 3-1-315, 1st Floor
Nimboliadda, Kachiguda, **Hyderabad** 500 027
email: ncbahydb@ncbapvtltd.com

GSS Shopping Complex, 1st Floor
Opposite College Post Office, Convent Road
Ernakulam 682 035
email: ncbaernk@ncbapvtltd.com

Radhamayee Complex (Second Floor)
Opp. SBI, Panbazar, Jaswanta Road
Guwahati 781 001
email: ncbaguwahati@ncbapvtltd.com

NUMERICAL METHODS PRACTICAL • Shyamapada Dutta and Harisankar Mallick

First Published: 1994
Reprinted: 2008, 2010, 2014

PUBLISHER AND TYPESETTER
New Central Book Agency (P) Ltd
8/1 Chintamoni Das Lane, Kolkata 700 009

PRINTER
New Central Book Agency (P) Ltd
Web-Offset Division, Dhulagarh, Sankrail, Howrah

COVER DESIGNER
Dulal Halder

COVER PRINTER
Biswanath Studio, Kolkata

PROJECT COORDINATORS
Prabhat Jas, Abhijit Sen

ISBN 978 81 7381 564 5

Price: ₹ 195.00

This Book is
Respectfully Dedicated
to the Hallowed Memory
of Our Grandfather
Late Jatindranath Mallick

Shyamapada
Harisankar

Syllabus

Burdwan University

NUMERICAL METHODS PRACTICAL

Marks 30

Sessional—Marks 10

Problem (Including VIVA)—Marks 20 (15 + 5)

Solving Equations for Real Roots by

(a) Regula Falsi

(b) Graphical

(c) Newton-Raphson Methods

To interpolate from a table using forward and backward formulae of Newton (use of table of at least six equidistant points).

Integration of a given function by

(a) Trapezoidal Rule and

(b) Simpson's One-third Rule.

Numerical Solution of a first order Differential Equation by Euler's Method with at least two steps from the initial value.

To fit a Straight line for a given data by the principle of least squares.

Preface

Dearth of books on Numerical Methods Practical by Indian authors has practically prompted us to endeavour at bringing out a suitable book to cater to the needs of the students of Mathematics of all the Indian Universities although the book is primarily meant for the BA and BSc (Honours) students of Burdwan University. To make this attempt a success we have consulted the master works of most of the celebrated foreign authors and the questions set on numerical methods in different years.

We take this opportunity to express our heartfelt gratitude to Dr K K Mandal, Dept of Mathematics, Burdwan University, but for whose help the book would not have seen the light of day. We are particularly indebted to Dr Santosh Kr Mallik, MS, FRCS, Head of the Dept of Surgery, Bankura Sammilani Medical College, whose prudent advice and encouragement has always been a constant source of inspiration to us. We shall also be failing in our duty if we do not express our sincere thanks to Smt Kakali Sadhu (Dutta), MSc (Maths), who helped us in many ways.

We shall consider our efforts fruitful if this book is of any help to the students. We shall deem it a privilege if we get constructive suggestions and criticisms from the teachers concerned for the betterment of this book in the subsequent editions.

15 April 1994

Shyamapada Dutta
Harisankar Mallick

Contents

Chapter 1

Interpolation

1. Newton's Formula for Forward Interpolation is

$$\begin{aligned}\phi(x) &= \phi(x_0 + hu)\\ &= y_0 + u\Delta y_0 + \frac{u(u-1)}{2!}\Delta^2 y_0 + \frac{u(u-1)(u-2)}{3!}\Delta^3 y_0 + \cdots\end{aligned}$$

where $u = \dfrac{x - x_0}{h}$.

The point x_0 is called the starting point of the formula.

1. $u, \dfrac{u(u-1)}{2!} \cdots$ are called co-efficients of the forward interpolation formula. It is used for interpolating at the beginning of the table.

2. Newton's formula for Backward Interpolation is

$$\begin{aligned}\phi(x) &= \phi(x_n + hu) = \psi(u)\\ &= y_n + u\Delta_1 y_n + \frac{u(u+1)}{2!}\Delta_2 y_n + \frac{u(u+1)(u+2)}{3!}\Delta_3 y_n + \cdots\end{aligned}$$

where $u = \dfrac{x - x_n}{h}$, or, $x = x_n + hu$.

This formula is used mainly for interpolating values of y near the end of a set of Tabular values and also for extrapolating values of y a short distance ahead of y_n.

How to constract the difference table?

(i) The zeros between the decimal point and first significant digit to its right are omitted. Thus, instead of writing .000652583 in problem 1 first number in column Δy we write simply 652583.

(ii) The number should he chosen so that the differences $\Delta^n y$ are practically constant.

Problem 1.1 **Date / /199**

Newton's Interpolation Formula

Find the value of $f(26.605)$, $f(26.592)$, $f(26.765)$, $f(26.804)$ by Newton's Interpolation formula.

X	$f(x)$
26.60	1.424881637
26.64	1.425534220
26.68	1.426185825
26.72	1.426836454
26.76	1.427486109
26.80	1.428134794

Difference Table

x	$y = f(x)$	Δy	$\Delta^2 y$	$\Delta^3 y$
26.60	1.424881637			
		652583		
26.64	1.425534220		−978	
		651605		2
26.68	1.426185825		−976	
		650629		2
26.72	1.426836454		−974	
		649655		4
26.76	1.427486109		−970	
		648685		
26.80	1.428134794			

Newton's forward interpolation formula is

$$f(x) = y_0 + u\Delta y_0 + \frac{u(u-1)}{\lfloor 2}\Delta^2 y_0 + \frac{u(u-1)(u-2)}{\lfloor 3}\Delta^3 y_0 + \cdots.$$

Here $x = 26.605, x_0 = 26.60, h = .04.$

$$\therefore \quad u = \frac{x - x_0}{h} = \frac{.005}{.040} = \frac{5}{40} = \frac{1}{8} = .125.$$

Coefficients	Multipliers	+ve Product	−ve Product
1	1.424881637	1.424881637	
.125	.000652583	.000081573	
−.0547	−.000000978	.0000000535	
.06836	.000000002	.0000000001	
	Total	1.4249632636	
	∴ f(26.605) =	1.424963263	

Now when $x = 26.592, x_0 = 26.60, h = .04.$

$$\therefore \quad u = \frac{x - x_0}{h} = \frac{26.592 - 26.60}{.04} = -.2.$$

Coefficients	Multipliers	+ve Product	−ve Product
1	1.424881637	1.424881637	
−.2	.000652583		.000130516
+.12	.000000978	.000000117	
−.088	.000000002		.0000000001
	Total	1.424881754	.0001305161
	Difference	1.42475123	

Ans. $f(26.592) = 1.424751238.$

Newton's Backward Interpolation formula is

$$y = y_n + u\Delta y_n + \frac{u(u+1)}{\underline{|2}}\Delta^2 y_n + \frac{u(u+1)(u+2)}{\underline{|3}}\Delta^3 y_n + \cdots.$$

Now when $x = \mathbf{26.765}, x_n = 26.80.$

$$\therefore \quad u = \frac{x - x_n}{h} = \frac{26.765 - 26.80}{.04} = -.875.$$

Coefficient	Multipliers	Positive term	Negative term
1	1.428134794	1.428134794	
−.875	.000648685		.0005675994
−.0547	−.000000970	.000000053	
	Total	1.428134847	.0005675994
	Difference	1.427567248	

Ans. $f(26.765) = 1.427567248$.

When $x = 26.804, x_n = 26.80, h = .04, u = .1$.

Coefficient	Multipliers	Positive term	Negative term
1	1.428134794	1.428134794	
.1	.000648685	.000064868	
.055	−.000000970		.000000053
	Total	1.428199662	.000000053
	Difference	1.428199609	

Ans. $f(26.804) = 1.428199609$.

Problem 1.2 **Date / /199**

From the following table determine $f(0.1245), f(0.1895), f(0.1175), f(0.1911)$

Using Data

x	$f(x)$
0.12	1.127497
0.13	1.138828
0.14	1.150274
0.15	1.161834
0.16	1.173511
0.17	1.185305
0.18	1.197217
0.19	1.209250

x	$f(x)$	Δy	$\Delta^2 y$	$\Delta^3 y$
0.12	1.127497			
		11331		
0.13	1.138828		115	
		11446		−1
0.14	1.150274		114	
		1156		3
0.15	1.161834		117	
		11677		0
0.16	1.173511		117	
		11794		1
0.17	1.185305		118	
		11912		3
0.18	1.197217		121	
		12033		
0.19	1.209250			

By Newton's Forward Interpolation Formula we get

$$g(u) = f(x) = y_0 + u\Delta y_0 + \frac{u(u-1)}{\underline{|2}}\Delta^2 y_0 + \cdots.$$

Here $u = \dfrac{x - x_0}{h}$, where $x = 0.1245, x_0 = 0.12, h = x_1 - x_0 = 0.01$.

$$\therefore \quad u = 0.45 \quad \text{and} \quad \frac{u(u-1)}{2} = -.12375.$$

Coefficient	Multiplier	(+)*ve* Product	(−)*ve* Product	Total
1	1.127497	1.127497		
0.45	0.011331	0.00509895		
−0.12375	0.000115		0.0000142312	
		1.13259595	0.0000143312	1.1325817188

Hence $f(0.1245) = 1.1325817188 = 1.13258172$ (up to 9 significant figures).

Here $u = \dfrac{x - x_o}{h}$, where $x = 0.1895, x_n = 0.19, h = 0.01$.

$$\therefore \quad u = -0.05 \quad \text{and} \quad \frac{u(u+1)}{2} = -0.02375.$$

Newton's Backward Interpolation Formula is

$$\phi(x) = \psi(x) = y_n + u\Delta_1 y_n + \frac{u(u+1)}{2}\Delta_2 y_n + \cdots.$$

Coefficient	Multiplier	(+)*ve* Product	(−)*ve* Product	Total
1	1.309250	1.209250		
−0.05	0.012033		0.00060165	
−0.02375	0.000121		0.00000287375	
		1.209250	0.00060452375	1.20864547625

Hence $f(0.1895) = 1.20864547625 = 1.20864548$ (up to 9 significant figure).

$$\phi(u) = y_o + u\Delta y_o + \frac{u(u-1)}{2}\Delta^2 y_o + \cdots.$$

Here $x = 0.1175, x_o = 0.12, h = x_o - x_o = 0.01, u = \dfrac{x - x_o}{h} = -0.25, \dfrac{u(u-1)}{2} = 0.15625.$

Here the function $f(0.1175)$ is outside the range of given values, is called extrapolation.

Coefficient	Multiplier	(+)*ve* Product	(−)*ve* Product	Total
1	1.127497	1.127437		
−0.25	0.011331		0.00283275	
0.15625	0.000115	0.0000179687		
		1.1275149687		1.1246822187

Hence $f(0.1175) = 1.1246822187 = 1.12468222$ (up to nine significant figure).

Here

$$\psi(u) = y_n + u\Delta_1 y_n + \frac{u(u+1)}{2}\Delta_2 y_n + \cdots$$

$x = 0.1911, x_n = 0.19, h = 0.01.$

$$\therefore \quad u = \frac{x - x_n}{h} = 0.11, \quad \frac{u(u+1)}{2} = 0.06105.$$

Coefficient	Multiplier	(+)ve Product	(−)ve Product	Total
1	1.209250	1.209250		
0.11	0.012033	0.00132363		
0.06105	0.000121	0.00000738705		
		1.21058101705		1.21058101705

Hence $f(0.1911) = 1.21058101705 = 1.21058102$ (up to nine significant figures).

Problem 1.3 **Date / /199**

Using Values

x	$f(x)$
1.00	1.1917780
1.10	1.1404073
1.20	1.0952444
1.30	1.0552215
1.40	1.0191588
1.50	0.9866783

Find the value of $f(0.98), f(1.03), f(1.42), f(1.55)$.

x	y	Δy	$\Delta^2 y$	$\Delta^3 y$	$\Delta^4 y$	$\Delta^5 y$
1.00	1.1917780					
		−513707				
1.10	1.1404073		62078			
		−451629		−11678		
1.20	1.0952444		50400		2880	
		−401.229		−8798		−862
1.30	1.0651215		41602		2018	
		−359627		−6780		
1.40	1.0191588		34822			
		−324805				
1.50	0.9866783					

Newton's forward formula is

$$f(x) = y_o + u_{c_1}\Delta y_o + u_{c_2}\Delta^2 y_o + u_{c_3}\Delta^3 y_o + \cdots.$$

Here $x = 0.98, x_o = 1.00, h = 0.10$.

$$\therefore \quad u = \frac{x - x_o}{h} = \frac{0.98 - 1.00}{0.10} = -0.2.$$

Coefficient	Multipliers	$-ve$ Product	$-ve$ Product
1	1.191778	1.191778	
−0.2	−0.0513707	0.0702742	
0.12	0.0062078	0.0007449	
−0.088	−0.0011678	0.0001028	
0.0704	0.000288	0.0000203	
−0.059136	−0.0000862	0.0000051	
		1.2029252	

Ans. $f(0.98) = 1.2029252$.

Newton's forward formula is

$$f(x) = y_o + u\Delta y_o + \frac{u(u-1)}{2}\Delta^2 y_o + \frac{u(u-1)(u-2)}{3.2}\Delta^2 y_o + \cdots$$

Here $x = 1.03; x_o = 1.00; h = 0.10$

$$\therefore \quad u = \frac{x - x_o}{h} = \frac{1.03 - 1.00}{0.10} = 0.3.$$

Coefficient	Multipliers	$-ve$ Product	$-ve$ Product
1	1.1917780	1.1917780	
0.3	−0.0513707		0.0154112
−0.105	0.0062078		0.0006518
0.0595	−0.0011678		0.0000695
−0.0401625	0.0002880		0.0000116
0.0297203	−0.0000862		0.0000026
			0.0161467

$$\therefore \quad f(1.03) = 1.1917780 - 0.0161467 = 1.1756313.$$

Newton's Backward formula is

$$f(x) = y_n + u\Delta y_n + \frac{u(u+1)}{\underline{|2}}\Delta^2 g_n + \frac{u(u+1)(u+2)}{\underline{|3}}\Delta^3 y_n + \cdots$$

Here $x = 1.55; x_n = 1.50$

$$\therefore \quad u = \frac{x - x_n}{h} - 1.55 - 1.510 = 0.5$$

Coefficient	Multipliers	$+ve$ Product	$-ve$ Product
1	0.9866783	0.9866783	
0.5	−0.0324805		0.0162403
0.375	0.0034822	0.0013058	
0.3125	−0.0006780		0.0002119
0.2734375	0.0002018	0.0000552	
0.2460937	−0.0000862		0.0000212
	Total	0.9880393	0.0164734

$$\therefore \quad f(1.55) = 0.9880393 - 0.0164734 = 0.9715659.$$

Here $x = 1.42$ $x_n = 1.50$

$$\therefore \quad u = \frac{x - x_n}{h} = \frac{1.42 - 1.50}{.10} = -0.8.$$

Coefficient Multipliers +*ve* Product −*ve* Product

Coefficient	Multipliers	+*ve* Product	−*ve* Product
1	0.9866783	0.9866783	
−0.8	−0.0324805	0.0259844	
−0.08	0.0034822		0.000278576
−0.032	−0.0006780	0.000021696	
−0.0176	0.0002018		0.0000035516
		1.012684396	0.000282127

$$\therefore \quad f(1.42) = 1.012402269.$$

Exercise 1

1. From the following table determine:

$$f(667), f(701), f(698)$$

X	$f(x)$
665	2.82282164
670	2.82607480
675	2.82930377
680	2.83250891
685	2.83569057
690	2.83884909
695	2.84198480
700	2.84509804

2. Find the value of y at $x = 4.189$ by Newton's interpulation formula

x	y
4.000	1.88649060
4.040	1.89100190
4.080	1.89551090
4.120	1.90001730
4.160	1.90452080
4.200	1.90902070

3. Find $f(.513)$, $f(.568)$ from the following table

x	$f(x)$
0.51	1.665291195
0.52	1.68202765
0.53	1.698932309
0.54	1.716006862
0.55	1.733253018
0.56	1.750672500
0.57	1.768267051

4. Find the value of $f(4.168)$

x	y
4.000	1.95424250
4.0400	1.95964050
4.0800	1.96502470
4.1200	1.97039520
4.1600	1.97515160
4.2000	1.98104340

5. Find the value of $f(32), f(54), f(29), f(57)$

x	y
30	0.500
35	0.573576436
40	0.642787609
45	0.707106781
50	0.766044443
55	0.819152044

6. Find the value of y when $x = 0.316$

From the following data

x	y
0.30	0.6179114
0.32	0.6255158
0.34	0.6330717
0.36	0.6405764
0.38	0.6480273
0.40	0.6554212

7. Find the value of $f(6.619), f(6.789)$

x	$f(x)$
6.60	0.531478917
6.64	0.526339277
6.68	0.521138083
6.72	0.515873843
6.76	0.510545010
6.80	0.505149978

8. Find $f(x)$ at $x = 3.912$ and at $x = 4.087$

x	$f(x)$
3.90	1.480150725
3.94	1.484635754
3.98	1.489120002
4.02	1.493603024
4.06	1.498084386
4.10	1.502563669

9. Calculate $f(1.151), f(1.156)$ & $f(1.182)$ from the following data.

x	$y = f(x)$
1.150	0.869565217
1.155	0.865800865
1.160	0.862068965
1.165	0.858369098
1.170	0.854700854
1.175	0.851063829
1.180	0.847457627
1.185	0.843881856

10. Find y for $x = 0.316$

x	y
0.30	0.6179114
0.32	0.6255158
0.34	0.6330717
0.36	0.6405764
0.38	0.6480273
0.40	0.6554212

11. Find $f(5.020), f(5.180)$

x	$f(x)$
5.000	1.609437912
5.040	1.617406082
5.080	1.625311262
5.120	1.633154439
5.160	1.640936579
5.200	1.648658626

12. Using the following Table find the value of $f(.7305), f(.7365)$

x	$f(x)$
.730	1.278494799
.731	1.279292024
.732	1.280090529
.733	1.280890313
.734	1.281691379
.735	1.282493726
.736	1.283297355
.737	1.284102268

13.

x	$f(x)$
1.50	0.405465108
1.51	0.412109650
1.52	0.418710334
1.53	0.425267735
1.54	0.431782416
1.55	0.438254930
1.56	0.444685821
1.57	0.451075619
1.58	0.457424847
1.59	0.463734016
1.60	0.470003629

From the above table find the values of

$$f(1.50911) f(1.50820) f(1.50253) f(1.59513) f(1.59575) f(1.59614) f(1.59729)$$

14. Find the value of $f(1.31)$&$f(1.39)$

x	$f(x)$
1.30	.1713686
1.32	.1669370
1.34	.1625551
1.36	.1582248
1.38	.1539483
1.40	.1497275

15. Find the value of $f(4.010)$ and $f(4.195)$ from the following Table.

x	$f(x)$
4.000	1.386294361
4.040	1.396244692
4.080	1.406096988
4.120	1.415853163
4.160	1.425515074
4.200	1.435084525

16. Function $f(x)$ is given by the Table. Using Newton's forward and Backward formula, find the values of these functions for the following values of the argument:

x	$f(x)$
1.50	1.144714243
1.51	1.147252420
1.52	1.149779416
1.53	1.152295353
1.54	1.154800350
1.55	1.157294527
1.56	1.159778000
1.57	1.162250881
1.58	1.164713286
1.59	1.167165320
1.60	1.169607095

Answers

1. $f(667) = 2.8241258$

$f(701) = 2.8457180$

$f(698) = 2.8438554.$

3. $f(.513) = 1.67029457$

$f(.568) = 1.764734052.$

4. $f(4.168) = 1.9768211264.$

5. $f(32) = 0.529919264$

$f(54) = 0.809016994$

$f(29) = 0.48480962$

$f(57) = 0.838670567.$

6. $y = 0.6239987.$

7. $f(6.619) = 0.52904517$

$f(6.789) = 0.506640305.$

8. $1.481496289, 1.501168155.$

9. $f(1.151) = 0.86880973,$

$f(1.156) = 0.865051903,$

$f(1.182) = 0.846023688.$

10. $y = 0.6239987.$

11. $f(5.020) = 1.613429934,$

$f(5.180) = 1.644805056.$

12. $f(.7305) = 1.278893252,$

$f(.7365) = 1.283699651.$

13. $f(1.50911) = 0.411520073$

$f(1.50820) = 0.410916886$

$f(1.50253) = 0.407150353$

$f(1.59513) = 0.466955237$

$f(1.59575) = 0.467343845$

$f(1.59614) = 0.467588214$

$f(1.59729) = 0.488308443.$

14. $f(1.31) = 0.1691468$

$f(1.39) = 0.1518308.$

15. $f(4.010) = 1.388791241$

$f(4.195) = 1.43389334.$

16. $f(1.50911) = 1.147026977$

$f(1.50820) = 1.146796377$

$f(1.50253) = 1.145357465$

$f(1.50192) = 1.145202446$

$f(1.59513) = 1.168419225$

$f(1.59575) = 1.168570587$

$f(1.59614) = 1.168665779$

$f(1.59728) = 1.168943942.$

for NOTES

for NOTES

for NOTES

for NOTES

Chapter 2

Central Difference Interpolation Formulae

STIRLING'S AND BESSEL'S INTERPOLATION FORMULAS

Newton's formulas for Forward and Backward Interpolation discussed in chapter 1 are suitable for interpolation near the beginning and end of a table of difference a formulas are suited. The most important central differences formula are:

(i) Stirling's Interpolation formula and

(ii) Bessel's Interpolation formula.

1. The stirling's formula is

$$y = y_0 + u\frac{\Delta y_1 + \Delta y_0}{2} + \frac{u^2}{2}\Delta^2 y_{-1} + \frac{u(u^2-1)}{3!}\frac{\Delta^3 y_{-2} + \Delta^3 y_{-1}}{2} + \cdots$$

where $u = \dfrac{x - x_0}{h}$.

2. The general form of Bessel's formula is

$$y = y_0 + u\Delta y_0 + \frac{u(u-1)}{2}\frac{\Delta^2 y_{-1} + \Delta^2 y_0}{2} + \frac{u\left(u-\frac{1}{2}\right)(u-1)}{3!}\Delta^3 y_{-1} + \cdots$$

where $u = \dfrac{x - x_0}{h}$.

A more symmetrical and convenient form of Bessel's formula is obtained by putting

$$u - \frac{1}{2} = v, \quad \text{or,} \quad u = v + \frac{1}{2}$$

$$\therefore \quad y = \frac{y_0 + y_1}{2} + v\Delta y_0 + \frac{\left(v^2 - \frac{1}{4}\right)}{2}\frac{\Delta^2 y_{-1} + \Delta^2 y_0}{2} + \frac{v\left(v^2 - \frac{1}{4}\right)}{3!}\Delta^3 y_{-1} + \cdots$$

Problem 2.1 **Date / /199**

Given

$$\log 410 = 2.6127839$$
$$\log 420 = 2.6232493$$
$$\log 430 = 2.6334685$$
$$\log 440 = 2.6434527$$
$$\log 450 = 2.6532125$$
$$\log 460 = 2.6627578$$

Find the values of $\log 4385$ by stirling's formula and also by Bessel's formula and compare the results obtained with its value given in the log-table.

x	$f(x)$	Δf	$\Delta^2 f$	$\Delta^3 f$	$\Delta^4 f$
410	2.6127839				
		104654			
420	2.6232493		−2462		
		102192		112	
430	2.6334685		−2350		−6
		99842		106	
440	2.6434527		−2244		−7
		97598		990	
450	2.6532125		−2145		
		95453			
460	2.6627578				

Now $\log 4385 = \log(438.5 \times 10) = \log 10 + \log 438.5 = 1 + \log 438.5$

We have stirling's formula

$$y = y_0 + u\frac{\Delta y_{-1} + \Delta y_0}{2} + \frac{u^2}{2}\Delta^2 y_{-1} + \frac{u(u^2-1)}{3!}\frac{\Delta^3 y_{-2} + \Delta^3 y_{-1}}{2} + \frac{u^2(u^2-1)}{4!}\Delta^4 y_{-2} + \cdots.$$

Here $x = 438.5,\ x_0 = 440,\ h = 10,\ u = \dfrac{x - x_0}{h} = -.15.$

Coefficient	Multipliers	$-ve$ Product	$-ve$ Product
1	2.6434527	2.6434527	
−.15	0.009872		0.0014808
.01125	−.0002244		0.0000025245
.0244375	0.0000548	0.0000013391	
−.0009164	−0.0000007	0.00000000	
		2.643454	0.0014833

$$\log 438.5 = 2.643454 - 0.0014833 = 2.6419707$$
$$\log 4385 = 1 + 2.6419707 = 3.6419707$$

We have by Bessel's formula

$$y = \frac{y_0 + y_1}{2} + v\Delta y_0 + \frac{\left(v^2 - \frac{1}{4}\right)}{2} \cdot \frac{\Delta^2 y_{-1} + \Delta^2 y_0}{2} + \frac{v\left(v^2 - \frac{1}{4}\right)}{3!}\Delta^3 y_{-2} + \cdots.$$

Here $x = 438.5,\ x_0 = 430,\ u = \dfrac{x - x_0}{h} = 0.85,\ v = u - \dfrac{1}{2} = .35.$

Coefficient	Multipliers	$+ve$ Product	$-ve$ Product
1	2.6384606	2.6384606	
.35	.0099842	.00349447	
−.06375	−.00002797	.0000017786	
−.0074375	.000000106		.00000000007
		2.6419567	

$$\log 4385 = 1 + 2.6419567 = 3.6419567.$$

We have

$$\log 4385 = 3.6419707 \quad \text{(By Stirling's formula)}$$
$$\log 4385 = 3.6419567 \quad \text{(By Bessel's formula)}$$
$$\log 4385 = 3.6419696 \quad \text{(By log-Table)}$$

Problem 2.2 Find $\cos 0.806595$ by Stirling formula, given

$$\cos 0.8050 = 0.693111235$$
$$\cos 0.8055 = 0.692750733$$
$$\cos 0.8060 = 0.692390058$$
$$\cos 0.8065 = 0.692029210$$
$$\cos 0.8070 = 0.691668188$$
$$\cos 0.8075 = 0.691306994$$
$$\cos 0.8080 = 0.690945627$$

x	y	Δy	$\Delta^2 y$
0.8050	0.693111235		
		−360502	
0.8055	0.692750733		−173
		−360675	
0.8060	0.692390058		−173
		−360848	
0.8065	0.692029210		−174
		−361022	
0.8070	0.691668188		−172
		−361194	
0.8075	0.691306994		−173
		−361367	
0.8080	0.690945627		

Stirling's formula is

$$y = y_0 + u\frac{\Delta y_{-1} + \Delta y_0}{2} + \frac{u^2}{2}\Delta^2 y_{-1} + \cdots$$

Here $x = 0.806595$, $x_0 = 0.8065$, $h = 0.0005$

$$\therefore \quad u = \frac{x - x_0}{h} = 0.19$$

Coefficient	Multipliers	+ve Product	−ve Product
1	.692029210	.692029210	
0.19	−.000360935		.0000685776500
0.01805	−.000000174		.0000000031407
		.692029210	.0000685807907

Ans. cos .806595 = 0.691960629

Exercise 2

1. Find the values of log 5.385 by stirling's formula and also by Bessel's formula and compare the results obtained with its value given in the log-table.

Given

log 5.10 = 0.707570176
log 5.20 = 0.716003343
log 5.30 = 0.724275869
log 5.40 = 0.732393759
log 5.50 = 0.740362689
log 5.60 = 0.748188027

2. Find the value of $f(1.7489), f(1.7486)$ by Stirling's formula and by Bessel's formula,

x	$f(x)$
1.72	1.311487705
1.73	1.315294644
1.74	1.319090596
1.75	1.322875656
1.76	1.326649916
1.77	1.330413470
1.78	1.334166406

3. Find the value of e^{-x} when $x = 1.2475$ using stirling's and Bessel's formula.

x	e^{-x}
1.22	0.295230166
1.23	0.292292577
1.24	0.289384217
1.25	0.286504796
1.26	0.283654026
1.27	0.280831621
1.28	0.278037300

4. Find the value of $\phi(0.5437)$ using Bessel's formula from the following table.

x	$\phi(x)$
0.51	0.5292437
0.52	0.5378987
0.53	0.5464641
0.54	0.5549392
0.55	0.5633233
0.56	0.5716157
0.57	0.5798158

5. Find the value of $\log_{10} 1445$ by stirling's formula and also by Bessel's formula and compare the results obtained with its given in the log-table.

x	$f(x)$
log 110	2.041392685
log 120	2.079181246
log 130	2.113943352
log 140	2.146128036
log 150	2.176091259
log 160	2.204119983
log 170	2.230448921

Answers

1. $\log 5.385 = 0.731185707$
2. $f(1.7489) = 1.322459829$
   $f(1.7486) = 1.322346399$
3. 0.287221954
4. $\phi(0.5437) = 0.5580520$
5. 3.159867847.

for NOTES

for NOTES

for NOTES

for NOTES

Chapter 3

Numerical Integration

In this chapter we shall find the approximate values of the definite integral form a given set of numerical values of the integrand. There are many such methods but we consider the following:

(i) Trapezoidal Rule and

(ii) Simpson's One-third Rule.

1. Trapezoidal Rule

(i) The Trapezoidal is

$$\int_a^b y\,dx = h\left[\frac{1}{2}y_0 + y_1 + y_2 + \cdots + y_{n-1} + \frac{1}{2}y_n\right];$$

(ii) It is applicable to any number of sub-intervals, whether even or odd;

(iii) The accuracy of the result can be improved by decreasing the interval h, i.e., increasing n, the number of sub-intervals;

(iv) Inherent error $= -\frac{nh^3}{12}f''(\xi) \quad x < \xi < b = -\frac{(b-a)}{12}h^2 f''(\xi)$.

Hence when $f(x)$ is a polynomial of the first degree, Trapezoidal rule gives the exact value of $\int_a^b f(x)dx$.

2. Simpson's one-third Rule :

(i) $\displaystyle\int_{x_0}^{x_0+nh} y\,dx = \frac{h}{3}[y_0 + 4(y_2 + y_3 + \cdots) + 2(y_2 + y_4 + \cdots) + y_n]$;

(ii) This rule can be applied when the range $[x_0, x_0 + nh]$ is divided into even number of sub-intervals;

(iii) Inherent error $= -\frac{b-a}{180}h^4 f^{IV}(\xi) \quad a < \xi < b$;

Hence when $f(x)$ is a polynomial of the first, second or third degree. Simpson's Rule gives the exact value of $\int_a^b f(x)dx$.

Problem 3.1 **Date / /199**

Evaluate $\int_{2.1}^{2.2} \frac{4dx}{x+\log_{10}(5x^2+4x+7)}$ by Trapezoidal rule taking $n = 10$ e.g., 11 ordinates correct up to 7 significant figures.

Trapezoidal rule is $\int_{x_o}^{x_n} ydx = h\left[\frac{1}{2}(y_o + y_n) + y_1 + y_2 + \cdots + y_{n-1}\right]$

$$h = \frac{2.2-2.1}{10} = \frac{.1}{10} = .01.$$

x	$f(x)$	C_n	$C_n y_n$
2.10	1.088894096	$\frac{1}{2}$	0.544447048
2.11	1.085085013	1	1.085085013
2.12	1.081304745	1	1.081304745
2.13	1.077552963	1	1.077552963
2.14	1.073829339	1	1.073829339
2.15	1.070133552	1	1.070133552
2.16	1.066465285	1	1.066465285
2.17	1.062824227	1	1.062824227
2.18	1.059210007	1	1.059210070
2.19	1.055622511	1	1.055622511
2.20	1.052061253	$\frac{1}{2}$	.526030626
			$10.70250532 \div 100 = .107025053$

$$\therefore \int_{2.1}^{2.2} ydx = 0.107025053 = 0.1070251 \quad \text{up to 7 significant figures.}$$

Problem 3.2 **Date / /199**

Find the value of $\int_2^3 \sqrt{\frac{x+10}{2x+1}} \log_{10} x dx$ by Trapezoidal rule taking $n = 6$, i.e., 7 ordinates. Correct to 5 significant figures.

Trapezoidal Rule is

$$\int_{x_0}^{x_0+xh} f(x)dx = \frac{h}{2}[y_0 + y_n + 2(y_1 + y_2 + \cdots + y_{n-1})], \quad y = \sqrt{\frac{x+10}{2x+1}}\log_{10} x$$

$$h = \frac{3-2}{6} = \frac{1}{6}.$$

x	$x+10$	$2x+1$	$\frac{x+10}{2x+1}$	$\sqrt{\frac{x+10}{2x+1}}$	$\log_{10} x$	y	C_n	$C_n y_n$
2	12	5	$\frac{12}{5}$	1.5491933	0.30103	0.4663536	$\frac{1}{2}$	0.2331768
$\frac{13}{6}$	$\frac{73}{6}$	$\frac{32}{6}$	$\frac{73}{32}$	1.5103807	0.3357921	0.5071739	1	0.5071739
$\frac{14}{6}$	$\frac{74}{6}$	$\frac{34}{6}$	$\frac{74}{34}$	1.4752866	0.3679767	0.542871	1	0.5428710
$\frac{15}{6}$	$\frac{75}{6}$	$\frac{36}{6}$	$\frac{75}{36}$	1.4433757	0.397940	0.5743769	1	0.5743769
$\frac{16}{6}$	$\frac{76}{6}$	$\frac{38}{6}$	$\frac{76}{38}$	1.4142136	0.4259687	0.6024107	1	0.6024107
$\frac{17}{6}$	$\frac{77}{6}$	$\frac{40}{6}$	$\frac{77}{40}$	1.3874437	0.4522976	0.6275374	1	0.6275374
$\frac{18}{6}$	$\frac{78}{6}$	$\frac{42}{6}$	$\frac{78}{42}$	1.3627703	0.4771212	0.6502066	$\frac{1}{2}$	0.3251033
							Total	3.4126500

$$\therefore \quad \int_2^3 \sqrt{\frac{x+10}{2x+1}}\log_{10} x\,dx = 3.41265 \times \frac{1}{6} = 0.568775 = 0.56978$$

(Correct up to 5 significant figure).

Problem 3.3 **Date / /199**

Evaluate

$$\int_{2.2}^{2.3} \frac{5dx}{x + \log_{10}(5x^2 + 2x + 8)}$$

by Simpson $\frac{1}{3}$ rule taking $n = 10$ i.e., 11 ordinates correct up to 7 significant figure.

Simpson $\frac{1}{3}$ rule is

$$\int_{x_0}^{x_n} ydx = \frac{h}{3}[y_0 + 4(y_1 + y_3 + \cdots) + 2(y_2 + y_4 + \cdots) + y_n]$$

$$h = \frac{2.3 - 2.2}{10} = \frac{.1}{10} = .01. \quad \frac{h}{3} = \frac{1}{300}.$$

x	$f(x)$	C_n	$C_n y_n$
2.20	1.328557228	1	1.328557228
2.21	1.324038402	4	5.296153608
2.22	1.319552577	2	2.639105154
2.23	1.315099393	4	5.260397572
2.24	1.310678492	2	2.621356984
2.25	1.306289520	4	5.225158080
2.26	1.301932132	2	2.603864264
2.27	1.297605985	4	5.190423940
2.28	1.293310742	2	2.586621484
2.29	1.289046068	4	5.156184272
2.30	1.284811638	1	1.284811638

Ans. $39.19263422 \div 3000 = 0.130642114$.

$$\therefore \int_{2.2}^{2.3} y\,dx = 0.130642114 = 0.1306421 \quad \text{(up to 7 significant figures).}$$

Problem 3.4 **Date / /199**

Compute by Simpson's $\frac{1}{3}$ rule the value of the integral $\int_2^3 \sqrt{\frac{x+5}{x+9}}\log_{10} x\,dx$ using 7 ordinates.

Simpson's $\frac{1}{3}$ rule is

$$\int_{x_0}^{x_0+nh} y\,dx = \frac{h}{3}[y_0 + 4(y_1 + y_3 + \cdots + y_{n-1}) + 2(y_2 + y_4 + \cdots + y_{n-2}) + y_n].$$

Here $y = \sqrt{\frac{x+5}{x+9}}\log_{10} x, h = \frac{1}{6}$.

x	$x+5$	$x+9$	$\frac{x+5}{x+9}$	$\sqrt{\frac{x+5}{x+9}}$	$\log_{10} x$	y	C_n	$C_n y_n$
2	7	11	$\frac{7}{11}$	0.797724	0.30103	0.2401388	1	0.2401388
$\frac{13}{6}$	$\frac{43}{6}$	$\frac{67}{6}$	$\frac{43}{67}$	0.8011186	0.3357921	0.2690093	4	1.0760372
$\frac{14}{6}$	$\frac{44}{6}$	$\frac{68}{6}$	$\frac{44}{68}$	0.8043996	0.3679767	0.2960003	2	0.5920006
$\frac{15}{6}$	$\frac{45}{6}$	$\frac{69}{6}$	$\frac{45}{69}$	0.8075728	0.39794	0.3213655	4	1.2854620
$\frac{16}{6}$	$\frac{46}{6}$	$\frac{70}{6}$	$\frac{46}{70}$	0.8106434	0.4259687	0.3453087	2	0.6906174
$\frac{17}{6}$	$\frac{47}{6}$	$\frac{71}{6}$	$\frac{47}{71}$	0.8136165	0.4522976	0.3679967	4	1.4719868
$\frac{18}{6}$	$\frac{48}{6}$	$\frac{72}{6}$	$\frac{48}{72}$	0.8164965	0.4771212	0.3895677	71	0.3895677
							Total	5.7458105

$$\therefore \int_2^3 \sqrt{\frac{x+5}{x+9}} \log_{10} x dx = 5.7458105 \times \frac{1}{18} = 0.3192116.$$

Problem 3.5 **Date / /199**

Evaluate $\int_4^{5.2} \frac{\log_e x}{1+x^2} dx$ by Simpson's $\frac{1}{3}$ rule, taking 12 equal intervals. Give the result correct to 6 significant figures.

Simpson's $\frac{1}{3}$ rule is

$$\int_{x_0}^{x_n} y dx = \frac{h}{3}[y_0 + 4(y_1 + y_3 + y_5 + \cdots) + 2(y_2 + y_4 + \cdots) + y_n]$$

where $h = \frac{x_n - x_0}{n}$.

Here

$$\left.\begin{aligned} x_n &= 5.2 \\ x_0 &= 4 \end{aligned}\right\} n = 12 \quad \therefore \quad h = \frac{5.2-4}{12} = 0.1.$$

Here $f(x) = y = \frac{\log_e x}{1+x^2} = \frac{\log_e 10 \times \log_{10} x}{1+x^2}.$

x	y_n	C_n	$C_n y_n$
4.0	0.0815467	1	0.0815467
4.1	0.0792244	4	0.3168976
4.2	0.0769895	2	0.1539790
4.3	0.0748392	4	0.2993568
4.4	0.0727704	2	0.1455408
4.5	0.0707801	4	0.2831204
4.6	0.0688654	2	0.1377308
4.7	0.0670231	4	0.2680924
4.8	0.0652502	2	0.1305004
4.9	0.0635440	4	0.2541760
5.0	0.0619015	2	0.1238030
5.1	0.0603199	4	0.2472796
5.2	0.0587966	1	0.0587966
			2.4948201

$$\therefore \int_4^{5.2} \frac{\log_e x}{1+x^2}dx = \frac{h}{3} \times 2.49482 = \frac{0.1}{3} \times 2.49482 = 0.0333333 \times 2.49482$$
$$= 0.0831606 \quad \text{(Correct up to 6 significant figure).}$$

Exercise 3

1. Evaluate

(i) $\int_0^1 \frac{1-x}{1+x}dx$; (ii) $\int_0^5 \log(1+x^2)dx$

by Simpson's $\frac{1}{3}$rd Rule and by Trapezoidal Rule with 10 sub-intervals and compare the results.

2. Evaluate by Simpson's $\frac{1}{3}$rd rule.

(a) $\int_{3.2}^{3.3} \frac{3x+7}{4x+\sqrt{3x^2+9x+8}}dx$ Taking 10 sub-internals up to 7 decimal places.

(b) Evaluate $\int_{3.2}^{3.3} \frac{(4x+4)dx}{x+\log_{10}(7x^2+5.5)}$ by Simpson's $\frac{1}{3}$rd rule and by Trapezoidal rule with 10 sub-intervals.

(c) Evaluate by Simpson $\frac{1}{3}$rd rule. $\int_0^{\frac{\pi}{10}} \frac{dx}{\sqrt{1-\frac{1}{2}\cos^2 x}}$ taking 12 equal intervals correct up to 6 significant figures.

3. $\int_2^3 \sqrt{\frac{x+5}{x+9}} \log_{10} x dx.$

Evaluate by Simpson's $\frac{1}{3}$rd rule using 7 ordinates.

4. Evaluate by Trapezoidal rule

(a) $\int_2^3 \sqrt{\frac{x+10}{2x+1}} \log_{10} x dx$ (Taking 7 ordinates, correct to 5 significant figures.)

(b) $\int_3^4 \sqrt{\frac{x+2}{x+1}} \log_{10} x dx (n=10)$

5. Taking 10 equal sub-intervals obtain by Simpsons $\frac{1}{3}$ rule the value of $\int_0^1 \frac{\log_e(1+x^2)}{1+x^2} dx.$

6. Evaluate by Simpson's $\frac{1}{3}$ rule $\int_5^{6.2} \log_e x dx$ (taking 7 equidistance ordinates)

7. Evaluate $\int_0^1 \frac{\log_e(1+x)}{1+x^2} dx$ (by Simpson's $\frac{1}{3}$ rule)

8. Evaluate $\int_0^{0.5} \frac{dx}{\sqrt{(1-x^2)(1-0.25x^2)}}$ (by Simpsons $\frac{1}{3}$ rule)

9. Evaluate $\int_0^{0.5} \frac{dx}{\sqrt{(1-x^2)(1-0.75x^2)}}$

10. Evaluate $\int_0^{0.5} \sqrt{\frac{1-0.75x^2}{1-x^2}} dx.$

11. Evaluate $\int_0^1 \frac{In(1+\sqrt{x})}{\sqrt[3]{x}} dx.$

12. Solve by Trapezoidal rule $\int_{2.1}^{2.2} \frac{4dx}{x+\log_{10}(7x^2+2x+7)}.$

13. Evaluate $\int_2^3 \sqrt{\frac{3x+1}{2x+1}} \log_{10} x dx$ by Simpson's $\frac{1}{3}$ rule taking $n = 10$, i.e., 11 ordinate correct up to 5 significant figures.

14. Find the value of $\int_{3.2}^{3.3} \frac{x+1}{x+\log_{10}(x^2+5.5)} dx.$

Answers

1. (i) 0.38629; (ii) 0.38754. **2.** (c) 0.437416. **3.** 0.3192113. **4.** (a) 0.568775, (b) 0.59983. **5.** 0.1730. **6.** 2.0650160. **7.** 0.2712826. **8.** 0.5294. **9.** 0.5422. **10.** 0.5061. **11.** 0.684. **12.** 0.1055512. **13.** 0.47020. **14.** (Try yourself).

for NOTES

for NOTES

for NOTES

for NOTES

Chapter 4

The Newton-Raphson Method

When the derivative of $f(x)$ is a simple expression and easily determine, the real root of $f(x) = 0$ can be found by this process.

The Newton-Raphson formula is

$$x_{n+1} = x_n - \frac{f(x_n)}{f'(x_n)}, \quad n = 0, 1, 2, \cdots$$

Provided $f'(x_n) \neq 0$.

Note

1. The method fails when $f'(x) = 0$.
2. When $f'(x) = 0$ Regula Falsi method may be used.
3. The sufficient (not necessary) condition of convergence of N-R method is

$$|f(x)f''(x)| < |f(x)|^2.$$

Working Rule

1. First find an approximate value x_0 of the root.
2. Find $f'(x)$.
3. Then constract columns of $x f(x), f'(x)$ and h, compute h correct to as many significant figures as there are zeros after the decimal points.

Problem 4.1 Date / /199

Find by Newton-Raphson Method the root of $x^2 - x\log_{10} x = 13$, upto 7 significant figures.

$$f(x) = x^2 - x\log_{10} x - 13$$
$$f'(x) = 2x - \log_{10} x - \log_{10} e.$$

x	$f(x)$	$f'(x)$	$h = -\dfrac{f(x)}{f'(x)}$
0	-13		
1	-12		
2	-9.60206		
3	-5.4313637	5.0885843	1.0673624
4	0.5917601	6.9636456	-0.0849784
3.9	-0.0951519	6.7746409	0.0140453
3.91	-0.0273111	6.7935288	0.0402016
3.914	-0.0001218	6.8010847	0.0000179089

$\therefore$ The required root $= 3.914 + 0.0000179089 = 3.9140179089 = 3.914018$ (up to 7 significant figures).

Problem 4.2 **Date / /199**

Find by Newton-Raphson Method the root of the Equation

$$f(x) = x^2 + x - 3\log_{10}(95x^2 + 5x + 4) = 0$$

bying between 1 and [illegible] correct up to 7 significant figures.

$$f'(x) = 2x + 1 - \frac{3(190x + 5)\log_{10} e}{95x^2 + 5x + 4}$$

x	$f(x)$
1	-4.0510999
2	-1.7864886
3	4.551379

$\therefore$ The root lies between 2 and 3.

x	$f(x)$	$f'(x)$	$h = -\frac{f(x)}{f'(x)}$
2	−1.7864886	3.7268778	−0.4793526
2.4	−0.0919878	4.7335417	−.0194331
2.41	−.04453090	4.75785854	−.00935944
2.419	−.0016181338	4.77971388	−.0003517724
2.4193	−.000177791	4.78044915	−.00003719127
2.41933	−.000034376	4.780514717	−.000007190857478

∴ The required root = 2.41933 + .0000071908 = 2.419337 (up to 7 significant figures.)

Exercise 4

1. Solve by Newton-Raphson method $x^2 + x\log_{10}(x+3) - 5 = 0$.
2. Solve by Newton-Raphson method
   (i) $f(x) = x^2 + 6x - \log_e(x^2 + 9x + 8)$;
   (ii) $15x = x\log_{10} x + 30$. Up to 5 sig. fig.
   (iii) $x + \ln x = 2$. Up to 5 decimal places.
   (iv) $f(x) = x + e^x - 9.4$. Up to 6 decimal places.
   (v) $x^2 - x\log_{10} x = 13$. Up to 7 sig fig.
   (vi) $x^2 - 5x + 4 + x^2\log_{10} x = 0$
   (vii) $3x - \cos x - 1 = 0$
   (viii) $10^x + x - 4 = 0$
   (ix) $2x - 3\sin x - 5 = 0$
   (x) $x^2 + x - 3\log_{10}(79x^2 + 4x + 5) = 0$ root lies between 2 and 3.

Answers

2. (ii) 2.0422198; (iii) 1.55715; (iv) 2.001304; (v) 3.9140181; (viii) 0.53918 (approx); (ix) 2.883238.

for NOTES

for NOTES

for NOTES

for NOTES

Chapter 5

Regula-Falsi Method

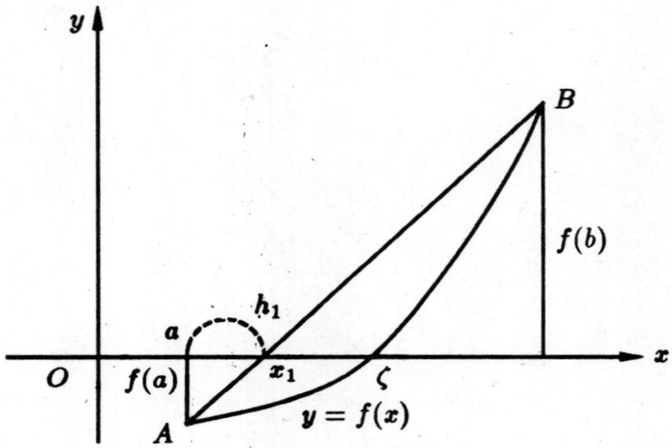

The formula for Regula-Falsi Method is

$$x = x_1 + h = x_1 + \frac{(x_2 - x_1)|y_1|}{|y_1| + |y_2|}$$

where $h = \dfrac{(x_2 - x_1)|y_1|}{|y_1| + |y_2|}$. (2)

x is not the true value of the root but it is merely a claser approximation to the true value of the root. Since the graph $y = f(x)$ is not a perfectly straight line between the two points.

For the practical purpose we compute a short table of corresponding values of x and $f(x)$ for equidistant values of x - units, tenths, hundreds etc. Then by (2) we compute corrections to be applied to the previously obtained approximate values.

Problem 5.1 **Date / /199**

Find the real of $2x - \log_{10} x = 7$ by Regula-Falsi method the root lying between 3 and 4 up to five decimal places.

Here $f(x) = 2x - \log_{10} x - 7 = 0, h = \dfrac{x_2 - x_1}{|y_1| + |y_2|}|y_1|$.

	x	$f(x)$	Remarks
	0	$-\infty$	
	1	-5	
	2	-3.3010	
	3	-1.4771212	One root lies between 3 and 4
	4	0.3979401	and $h_1 = 0.7$
1st approx	3.7	-0.1682017	The root lies between 3.7 and
	3.8	0.0202164	3.8 and $h_2 = 0.08$
2nd approx	3.78	-0.0174918	The root lies between 3.78
	3.79	0.0013608	and 3.79 and $h_3 = 0.009$
3rd approx	3.789	-0.0005246	The root lies between 3.789
	3.790	0.0013608	and 3.790 and $h_4 = 0.0002$
4th approx	3.7892	-0.0001475	The root lies between 3.7892
	3.7893	0.000041	and 3.7893 and $h_5 = 0.0000782078$

Hence the required root is $x = 3.7892 + 0.0000782078 = 3.78927$ [up to correct six significant figures]

Problem 5.2 **Date / /199**

Using Regula-Falsi method find the root of

$$\log_{10} x - 2x^2 + 8 = 0 \quad \text{correct up to 5 places of decimal}$$

Here $f(x) = \log_{10} x - 2x^2 + 8$

$$h = \frac{x_2 - x_1}{|f(x_1)| + |f(x_2)|} |f(x_1)|.$$

	x	$f(x)$	Remark
	0	$-\infty$	
	1	6	
	2	0.3010300	one root lies between
	3	−9.5228787	2 & 3 and $h = 0.03$
			$x_1 = 2 + .03 = 2.03$
1st	2.03	0.0656960	The root lies between
approx	2.04	−0.0135698	2.03 & 2.04 and
			$h = 0.0082 = 0.008$
2nd	2.038	0.0023162	The root lies between
approx	2.039	−0.0056248	2.038 & 2.039 and
			$h = 0.00029$
3rd	2.0382	0.00072832	The root lies between
approx	2.0383	−0.00006568	2.0382 & 2.0383
			and $h = .000091$
4th	2.03829	0.0000138	The root lies between
approx	2.0383	−0.00006568	2.03829 & 2.0383
			$h = .000001736$

So the required root is 2.03829117 = 2.03829 (up to 5 places of decimals)

Exercise 5

1. Solve by Regula-Falsi method the equation $x^2 + x\log_{10}(x+8) - 5 = 0$ up to 6 significant figures.

2. Solve by Regula-Falsi method the equation $x^2 + x\log_{10}(x+7) - 5 = 0$.

Ans. 1.812897.

3. Solve by Regula-Falsi method $4x^2 - \sqrt{x^2 + 2x + 12.5} =$ [0.991554] up to 6 decimal places.

4. Solve by Regula-Falsi method up to 6 significant figures the following equations

(i) $x \log_{10} x - 7x^2 + 35 = 0$. [2.2615256318]

(ii) $2x - \log_{10} x = 7$. [3.7893]

(iii) $6x - \log_{10}(x^2 + 8x + 16) = 0$. [0.208026]

(iv) $x + \ln x = 2$. [1.55715]

(v) $x^2 - 5 \log_{10}(5x^2 + 2x + 3) = 0$. [2.92694]

5. Find the smallest positive real root of $xe^x - 2 = 0$ [0.8526] up to 4 significant figures.

Answers

2. 1.812897; **3.** 0.991554; **4.** **(i)** 2.2615256318; **(ii)** 3.7893; **(iii)** 0.208026; **(iv)** 1.55715; **(v)** 2.92694; **5.** 0.8526.

for NOTES

for NOTES

for NOTES

for NOTES

Chapter 6

Euler's Method

Let us consider the differential Equation $\frac{dy}{dx} = f(xy)$ with the initial condition $y(x_\circ) = y_\circ$. We constract a table of y for given steps of h in x Eules's formula is

$$y_{n+1} = y_n + hf(x_n, y_n) \quad n = 0, 1, 2, \cdots$$

Draw backs of the method.

(i) In case h is small, this method is too slow. However if h is large this method is inaccurate.

Problem 6.1 Date / /199

Find by Euler's method, the value of y at $x = 1.12$ from the differential Equation

$$\frac{dy}{dx} = \frac{\log_{10}(x^2 + 2x + 4)}{x + y^2} \qquad y_{n+1} = y_n + hf(x_n y_n)$$

$x_\circ = 1.11, y_\circ = 3.0$ and $h = .001$.

n	x_n	y_n	$hf(x_n y_n)$	$y_n + hf(x_n y_n)$
0	1.110	3.0	0.000086278	3.000086278
1	1.111	3.00086278	0.00008629	3.000172568
2	1.112	3.000172568	0.000086301	3.00025887
3	1.113	3.000258870	0.000086312	3.000345183
4	1.114	3.000345183	0.000086324	3.000431507
5	1.115	3.000431507	0.000086335	3.000517843
6	1.116	3.000517843	0.000086346	3.000604190
7	1.117	3.000604190	0.000086358	3.000690548
8	1.118	3.000690548	0.000086369	3.000776918
9	1.119	3.000776918	0.000086380	3.000863299
10	1.120	3.000863229		

Ans. $\therefore \; y = 3.000863229$ at $x = 1.120$

Problem 6.2 **Date / /199**

Use Euler's Method with $h = 0.1$ to find the solution of the equation $\frac{dy}{dx} = x + y$ with initial condition $y = 1$ for $x = 0$ in the range $0 \le x < 1$.

n	x_n	y_n	$y_n + x_n$	$h(h_n + x_n)$	$y_{n+1} = y_n + hf(x_n y_n)$
0	0	1	1.00	0.1	1.10
1	0.10	1.1	1.20	0.12	1.22
2	0.20	1.22	1.42	0.142	1.3620
3	0.30	1.362	1.662	0.1662	1.5282
4	0.40	1.5282	1.9282	0.19282	1.72102
5	0.50	1.72102	2.22102	0.222102	1.943122
6	0.60	1.943122	2.543122	0.2543122	2.1974342
7	0.70	2.1974342	2.8974342	0.28974342	2.4871776
8	0.80	2.4871776	3.2871776	0.32871776	2.81589536
9	0.90	2.81589536	3.71589536	0.371589536	3.187484896
10	1.00	3.187484896			

Hence the solution of the differential equation is $y = 3.187484896$ when $x = 1$.

Exercise 6

1. Find by Euler's Method, the value of y at $x = 1.48$ from the differential equation.

$$\frac{dy}{dx} = \frac{\log_{10}(x^2 + 5y + 4)}{7x + y^2}$$

given $x_0 = 1.47, y_0 = 6.6, h = 0.001$.

2. Euler's method $\frac{dy}{dx} = \frac{7x + \log_{10}(70y + 55) - 1}{7y + 2}$.

   Given $x_0 = 1, y_0 = 1.5, h = .01$.

3. Find y at $x = 1.1$ where $\frac{dy}{dx} = \frac{4y + \log_{10}(2x^2 + 5.6)}{2y + x}$ given $h = .001, x_0 = 1, y_0 = 1$.

4. Solve by Euler's method $\frac{dy}{dx} = \frac{5y + \sqrt{x + 2.4}}{5x^2 + \log_{10}(2 + y)}$.

Given $y(1) = 2, h = .01$ find y at $x = 1.1$.

5. Solve by Euler's method $\frac{dy}{dx} = (10^{x+2} - 10^x)x^2$ find $y_{.25}$ given $y_0(0) = 1$, $h = .05$.

6. Solve by Euler's method

(i) $\frac{dy}{dx} = \frac{1}{2}xy \quad y(0) = 1, h = 0.1$ in (0, 1)

(ii) $\frac{dy}{dx} = 1 + xy^2 \quad y(0) = 0, h = 0.1$ in (0, 1)

7. Solve by Euler's method the following differential equations with the given condition $y(1) = 1$ in the interval $(1, 2)$:

(i) $\frac{dy}{dx} = -y^2 + \frac{0.05}{x^2}$;

(ii) $\frac{dy}{dx} = -y^2 + \frac{0.1}{x^2}$;

(iii) $\frac{dy}{dx} = -y^2 + \frac{0.15}{x^2}$.

Answers

1. 6.600295; **2.** 1.567065; **5.** 1.547332007; **6.** (i) 1.241794; (ii) 1.24547.

for NOTES

for NOTES

for NOTES

for NOTES

References

1. J. B. Scarborough : Numerical Mathematical Analysis, *Oxford & IBH Publishing Co.*, Sixth Edition, 1966.
2. F. B. Hildebrand : Introduction to Numerical Analysis, *Tata Mc-Graw Hill*, 2nd Edition.
3. A. Gupta and S. C. Bose : Introduction to Numerical Analysis, *A. P.*
4. Kalyan Kr. Mukherjee : Numerical Analysis, *New Central Book Agency (P) Ltd.*, 1st Edition, 1990.
5. B. P. Demidovich, F. A. Maron : Computational Mathematics, *Mir-Publication*.
6. H. C. Saxena : Finite Differences and Numerical Analysis, *S. Chand & Co.*, 1988.
7. Dutta and Jana : Introductory Numerical Analysis, *Shreedhar Prakashani*.

Chapter 7

Method of Fitting a Straight Line by the Principal of Least Square

The method of least square is based on the principal of least square of the residuals. Since the residuals are positive quantities, therefore the sum of square of the residuals will be minimum if the representative curve passes as closely as possible to all the points.

The method of fitting a straight line $y = a + bx$ to a set of n points

$$(x_1 y_1)(x_2 y_2)\ldots(x_n y_n)$$

by the least square principle.

Let us consider the st. line

$$y = a + bx. \tag{1}$$

We have to find the values of a and b such that the straight line (1) pass as near as possible to each of the points $(x_1, y_1)(x_2, y_2)\ldots(x_n, y_n)$. The equation will not, in general, be satisfied exactly by any of the n points. Putting in (1) the n pairs of values we get the following residual equations:

$$\nu_1 = a + bx_1 - y_1$$
$$\nu_2 = a + bx_2 - y_2$$
$$\nu_n = a + bx_n - y_n.$$

The principle of least squares says that the best values of the unknown constants a and b are those which make the sum of squares of the residuals a minimum, or

$$\Sigma\nu^2 = \nu_1^2 + \nu_2^2 + \ldots + \nu_n^2$$

must be a minimum.

Hence

$$f(a,b) = (a + bx_1 - y_1)^2 + (a + bx_2 - y_2)^2 + \ldots + (a + bx_n - y_n)^2$$

is to be a minimum.

The condition that $f(ab)$ will be maximum or minimum if

$$\frac{\partial f}{\partial a} = 0 \quad \text{and} \quad \frac{\partial f}{\partial b} = 0$$

$$\text{i.e.,} \quad \frac{\partial f}{\partial a} = 2(a + bx_1 - y_1) + 2(a + bx_2 - y_2) + \ldots + 2(a + bx_n - y_n) = 0$$

$$\frac{\partial f}{\partial b} = 2(a + bx_1 - y_1)x_1 + 2(a + bx_2 - y_2)x_2 + \ldots + 2(a + bx_n - y_n)x_n = 0.$$

Dividing by 2, we get the following normal equations:

$$(a + bx_1 - y_1) + (a + bx_2 - y_2) + \ldots + (a + bx_n - y_n) = 0$$

$$x_1(a + bx_1 - y_1) + x_2(a + bx_2 - y_2) + \ldots + x_n(a + bx_n - y_n) = 0$$

$$\text{or,} \quad na + b\Sigma x_i - \Sigma y_i = 0$$

$$a\Sigma x_i + b\Sigma x_i^2 - \Sigma x_i y_i = 0.$$

The normal eqautions are solved by Cramer's rule or by ordinary methods of algebra.

Here,

$$a = \frac{\begin{vmatrix} \Sigma x_i & \Sigma y_i \\ \Sigma x_i y_i & \Sigma x_i^2 \end{vmatrix}}{\begin{vmatrix} n & \Sigma x_i \\ \Sigma x_i & \Sigma x_i^2 \end{vmatrix}}, \quad b = \frac{\begin{vmatrix} n & \Sigma y_i \\ \Sigma x_1^2 & \Sigma x_i y_i \end{vmatrix}}{\begin{vmatrix} n & \Sigma x_i \\ \Sigma x_i & \Sigma x_i^2 \end{vmatrix}}.$$

The number of normal equation is always equal to the number of unknown constant to be determined. Putting these values of a and b in (1) we get the best fitted straight line by the method of least square.

Fit a straight line of the form $y = a_o + a_1 x$ where x is independent variable from the following data.

Problem 7.1 **Date / /199**

x	25	30	35	40	45	50	55	60
y	.54321	.45312	.31254	.12345	.32154	.27135	.12123	.12004

Table for Calculation

x	y	xy	x^2
25	.54321	13.58025	625
30	.45312	13.59360	900
35	.31254	10.93890	1225
40	.12345	4.93800	1600
45	.32154	14.46930	2025
50	.27135	13.56750	2500
55	.12123	6.66765	3025
60	.12004	7.20240	3600
340	2.26648	84.95760	15500

Normal Equations are

$$\Sigma y = na_o + a_1\Sigma x \quad 8a_o + 340a_1 = 2.26648$$

$$\Sigma xy = a_o\Sigma x + a_1\Sigma^2 \quad 340a_o + 15500a_1 = 85.95760$$

$$a_o = \frac{\begin{vmatrix} 2.26648 & 3.40 \\ 86.95760 & 15500 \end{vmatrix}}{\begin{vmatrix} 8 & 340 \\ 340 & 15500 \end{vmatrix}} = \frac{6244.856}{8400} = 0.7434352$$

$$a_1 = \frac{\begin{vmatrix} 8 & 2.2468 \\ 340 & 84.9576 \end{vmatrix}}{\begin{vmatrix} 8 & 340 \\ 340 & 15500 \end{vmatrix}} = \frac{-90.9425}{8400} = -.0108264.$$

∴ The required fitted straight line is

Ans. $y = 0.7434352 - 0.0108264x$

Problem 7.2 **Date / /199**

Fit a straight line of the form $y = a_o + a_1x$ where x is the independent variable from the following data.

x	25	30	35	40	45	50	55	60	65	70	75	80
y	45.071	50.041	43.550	51.791	52.771	60.606	61.610	63.219	72.912	73.112	75.601	59.001

x	y	$X = \frac{x-50}{5}$	$y = y - 50$	X^2	XY
25	45.071	−5	−4.9290	25	24.645
30	50.041	−4	0.0410	16	−0.1640
35	43.550	−3	−6.4500	9	19.350
40	51.791	−2	1.7910	4	−3.5820
45	52.771	−1	2.7710	1	−2.7710
50	60.606	0	10.606	0	0
55	61.610	1	11.610	1	11.610
60	63.219	2	13.219	4	26.438
65	72.912	3	22.912	9	68.736
70	73.112	4	23.112	16	92.448
75	75.601	5	25.601	25	128.005
80	59.001	6	9.0010	36	54.006
		$\Sigma X = 6$	$\Sigma Y = 109.285$	$\Sigma x^2 = 146$	$\Sigma XY = 418.721$

Problem 7.3 **Date / /199**

The normal Equations are

$$\Sigma Y = \Sigma A + \Sigma BX$$
$$\Sigma XY = A\Sigma X + B\Sigma X^2$$

or, $12A + 68 = 109.285$

$6A + 1468 = 418.721.$

$$A = \frac{\begin{vmatrix} 109.285 & 6 \\ 418.721 & 146 \end{vmatrix}}{\begin{vmatrix} 12 & 6 \\ 6 & 146 \end{vmatrix}} = 7.8340815$$

$$B = \frac{\begin{vmatrix} 12 & 109.285 \\ 6 & 418.721 \end{vmatrix}}{\begin{vmatrix} 12 & 6 \\ 6 & 146 \end{vmatrix}} = 2.5460834.$$

The required fitted straight line is

$$y - 50 = 7.8340815 + 2.5460835 \times \frac{x - 50}{5}$$

$$\text{or,}\quad 5y = 250 + 39.170407 - 127.30017 + 2.5460834x$$

$$\text{or,}\quad y = 32.3740466 + 0.5092006x \quad \textbf{(Ans.)}$$

Exercise 7

1. Fit the straight line of the form $y = a + bx$ by the principle of least square from the following data.

(i)

x	1	2	3	4	5	6	7	8
y	1.3154	2.1046	2.8988	3.6830	4.4722	5.2614	6.0506	6.8398

(ii)

x	y
1.46	3.336401
1.47	3.340789
1.49	3.349638
1.52	3.363090
1.56	3.381357
1.61	3.404713
1.67	3.433493
1.74	3.468079

(iii)

x	y
1.47	3.649897
1.48	3.657970
1.50	3.674227
1.53	3.698884
1.57	3.732257
1.62	3.774750
1.68	3.826844
1.75	3.889079

(iv) Fit a straight of the form $y = a_o + a_1 x$ from the following data

x	y
1.00	−3.000000
1.01	−3.009699
1.03	−3.027273
1.06	−3.048984
1.10	−3.069000
1.15	−3.079125
1.21	−3.068439
1.28	−3.022848

(v) Suppose that the following empirical data are available

x	1.36	1.49	1.73	1.81	1.95	2.16	2.28	2.48
$f(x)$	14.094	15.069	16.844	17.378	18.435	19.949	20.963	22.495

Fit the straight line of the form $y = a + bx$, by the method of the principle of least square.

2. Apply the principle of least square to fit a straight line in the form $y = a + bx$ by the following data.

x	4	6	8	10	12
y	13.72	12.90	12.01	11.14	10.31

3. Fit a straight line of the form $y = a + bx$ by the method of least sqaure from the following data.

x	y
10	3.0041
20	6.3172
30	9.6279
40	12.9368
50	16.2483
60	19.5613
70	22.8717
80	26.1801

Answers

1. **(i)** $y = 0.5262 + 0.7892x$; **(iv)** $y = 2.897842 - 0.219251x$. **2.** $y = 15.448 + 10.085x$. **3.** $y = -0.3058 + 0.3311x$.

for NOTES

for NOTES

for NOTES

for NOTES

Chapter 8

Graphical Method

Graphical method is solution of an equation by repeated plotting on a Larger Scale, and has the advantage of giving a visual representation of the approximating process.

We first find the approximate value of the root by a graph or otherwise. Then plot on a large scale a small part of the graph of $y = f(x)$ for values of x near the desired root, so that one can see more clearly about where the graph crosses the x-axis. An additional figure of the root can be found from this graph. Then plot on a still larger scale a small part of the graph for values of x near the improved value of the root and continue the process in this way until the root has found to as many figures as desired.

Problem 8.1 **Date / /199**

Solve graphically the equation $2x - \log_{10} x = 7$ upto four decimal places.

$$f(x) = 2x - \log_{10} x - 7 = 0.$$

x	$2x$	$\log_{10} x$	$f(x)$
1	2	0	−5
2	4	0.3010	−3.301
3	6	0.4771212	−1.4771212
4	8	0.6020599	0.3979401

Hence the root lies between $x = 3$ and $x = 4$.

3.7	7.4	0.5682017	−0.1682017
3.8	7.6	0.5797836	0.0202164
3.9	7.8	0.5910646	0.2089354

Hence the root lies between $x = 3.7$ and 3.8.

3.78	7.56	0.5774918	−0.0174918
3.79	7.58	0.5786392	0.0013608
3.80	7.60		

Hence the root lies between $x = 3.78$ and 3.79.

3.7880	7.576	0.5784099	−0.0024099
3.7890	7.578	0.5785246	−0.0005246
3.7900	7.580	0.5786392	0.0013608

Hence the root lies between $x = 3.789$ & $x = 3.790$.

3.7892	7.5784	0.5785475	−0.0001475
3.7893	7.5786	0.5785589	0.0000411

Hence the required root lies between 3.7892 and 3.7893

Exercise

Sove graphically $f(x) = 2x - \log_{10} x - 7 = 0$. (Set in graph)

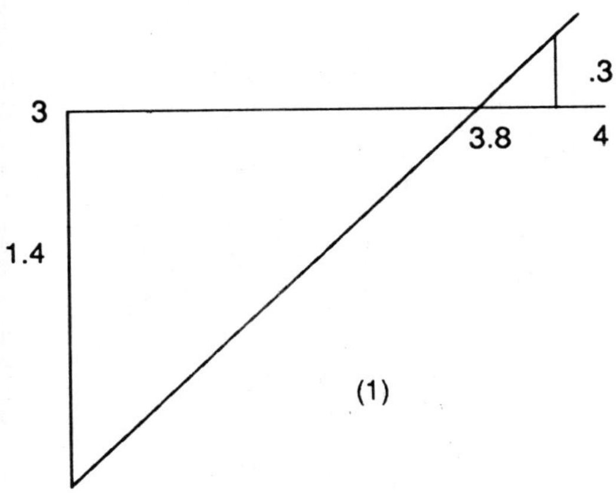

(1)

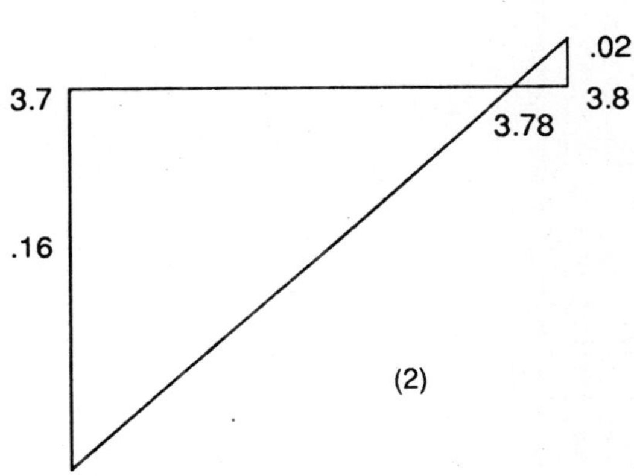

(2)

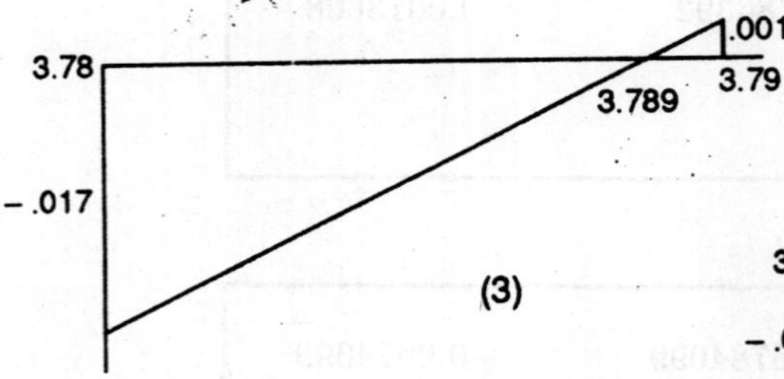

(3)

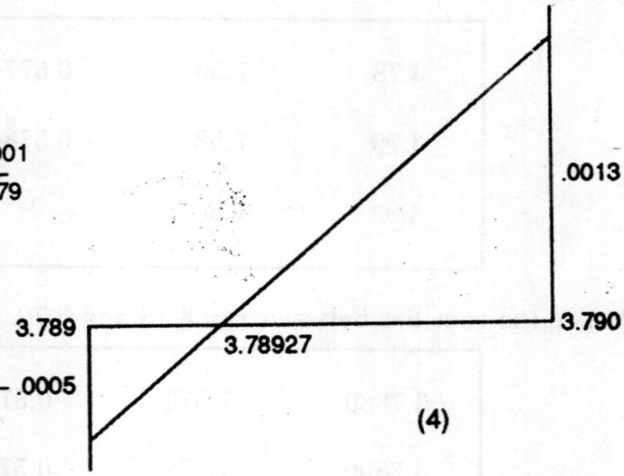

(4)

1. Solve by Graphical method

(a) $2x^2 + x\log_{10}(x+1) - 5 = 0$;
Up to 5 decimal places.

(b) $3x - \log_{10} x - 9 = 0$;

(c) $e^x - e^{-x} - 4x = 0$;

(d) $3x^2 + x\log_0(6x+10) - 43 = 0$;

(e) $xe^x = 4$;

(f) $x - \sin x = .02 = 0$;

(g) Find the positive root of $x - \cos\left(\frac{0.7854 - x\sqrt{1-x^2}}{1-2x^2}\right) = 0$.

Answers

1. (b) 3.506677. (c) 2.1773. (e) 1.202168. (f) 0.49526.

for NOTES

for NOTES

for NOTES

for NOTES